Sublimes cartes à jouer

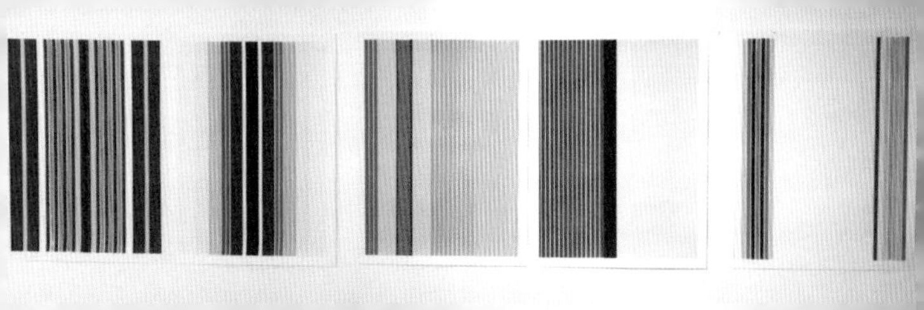

NOMA Ltd.

Appointment Only

NÒMA
Frame ©
Chain Stitch

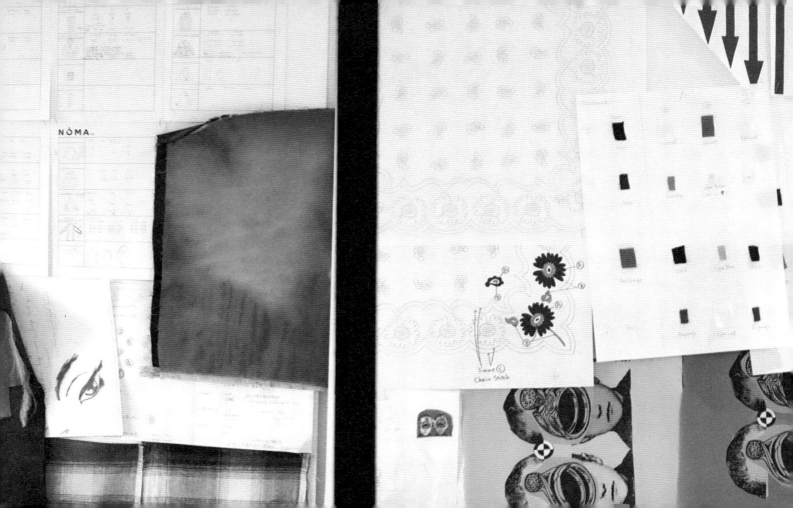

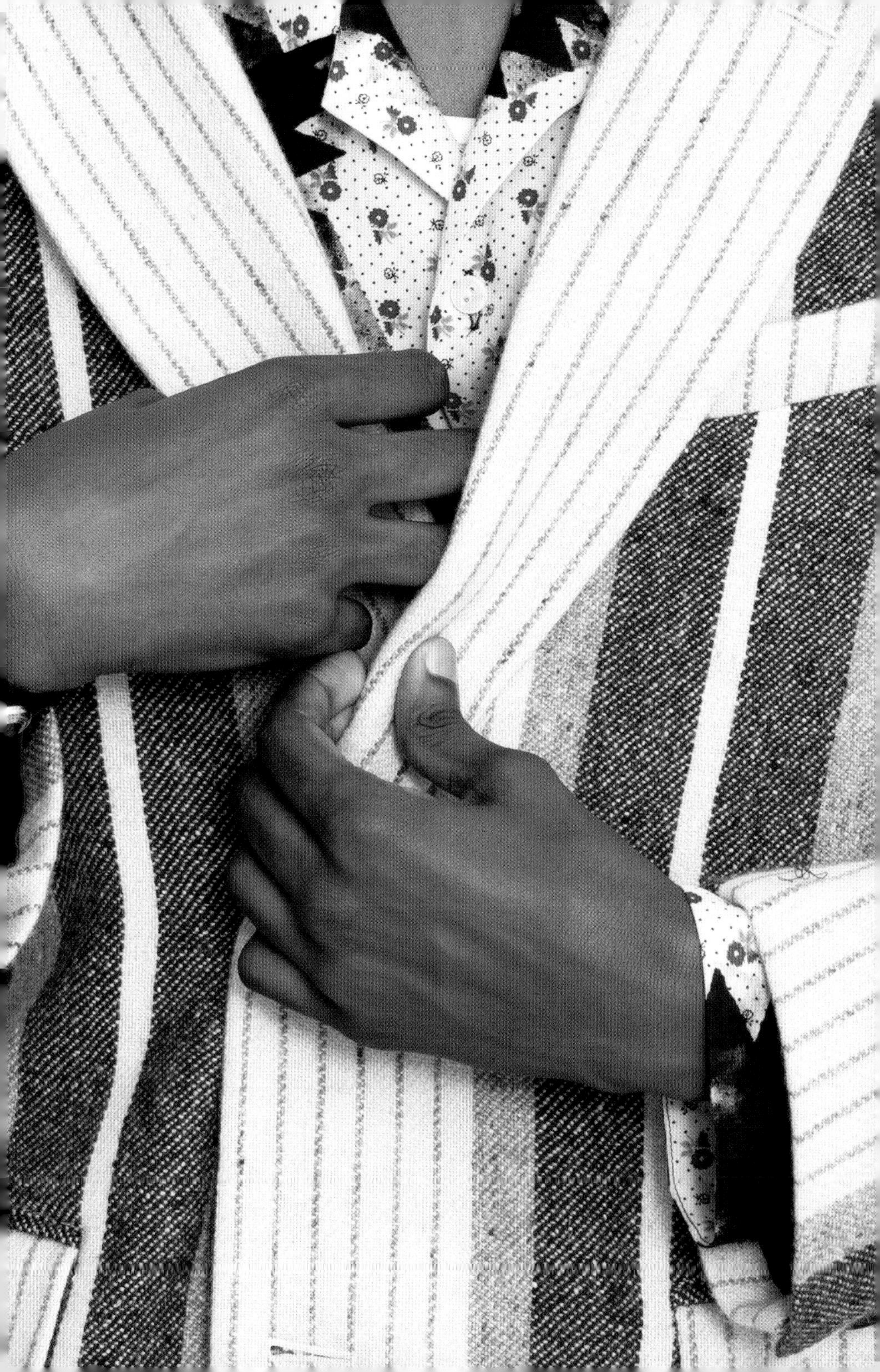

球団5月

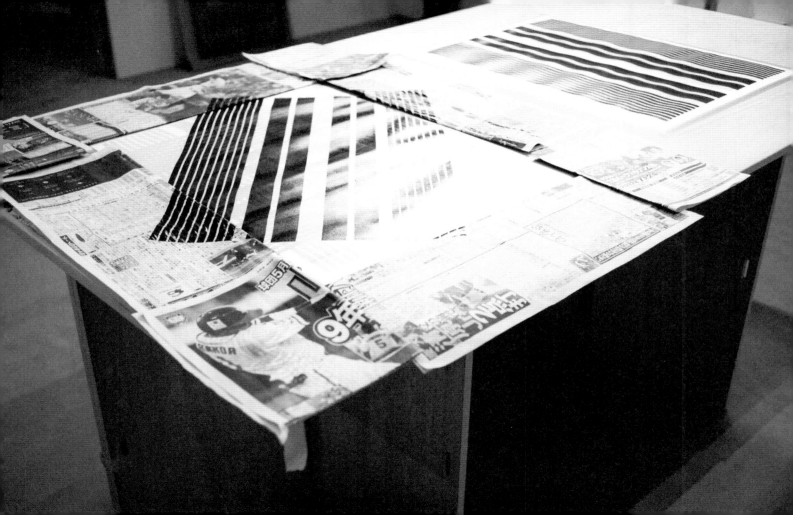

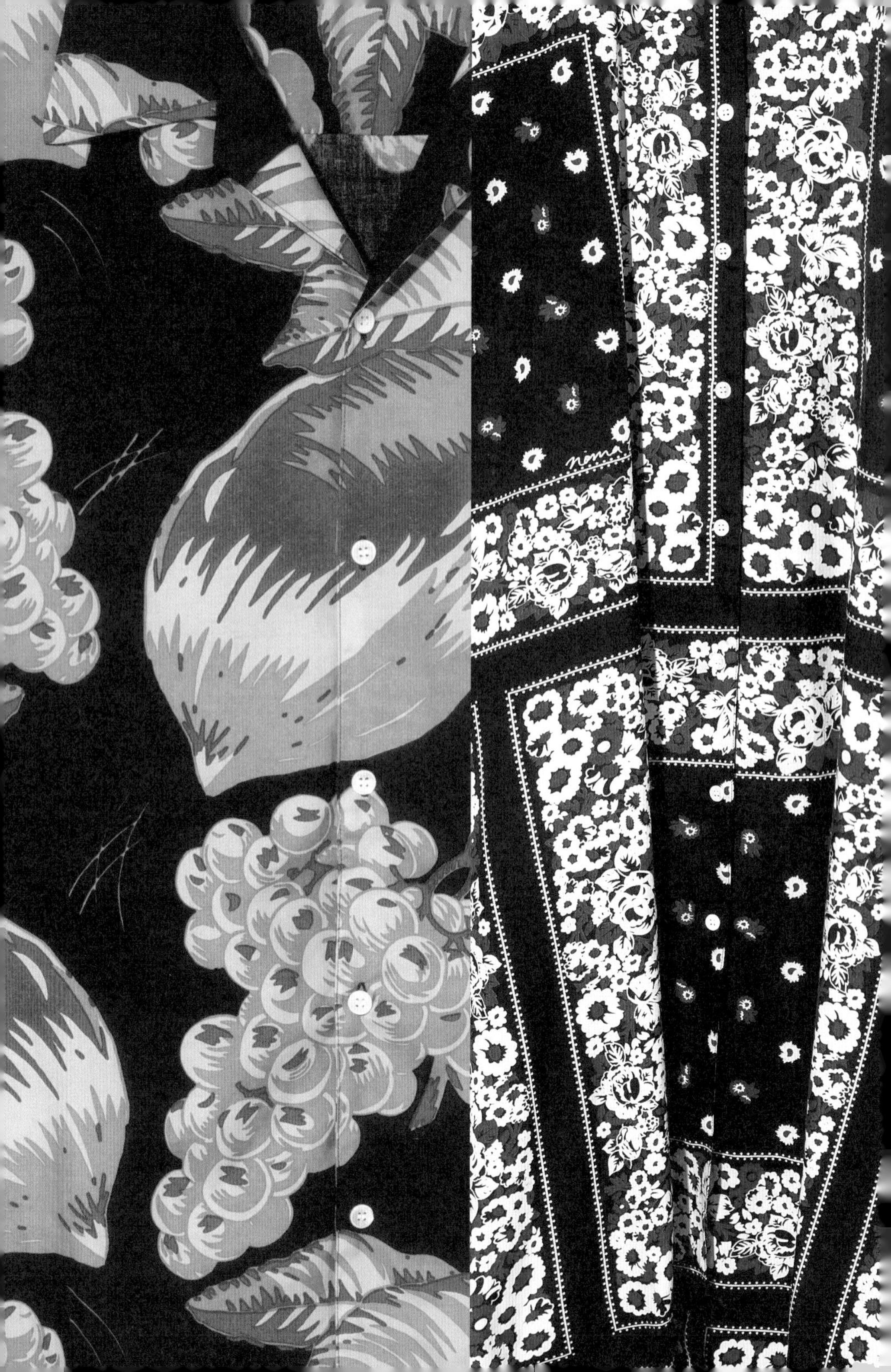

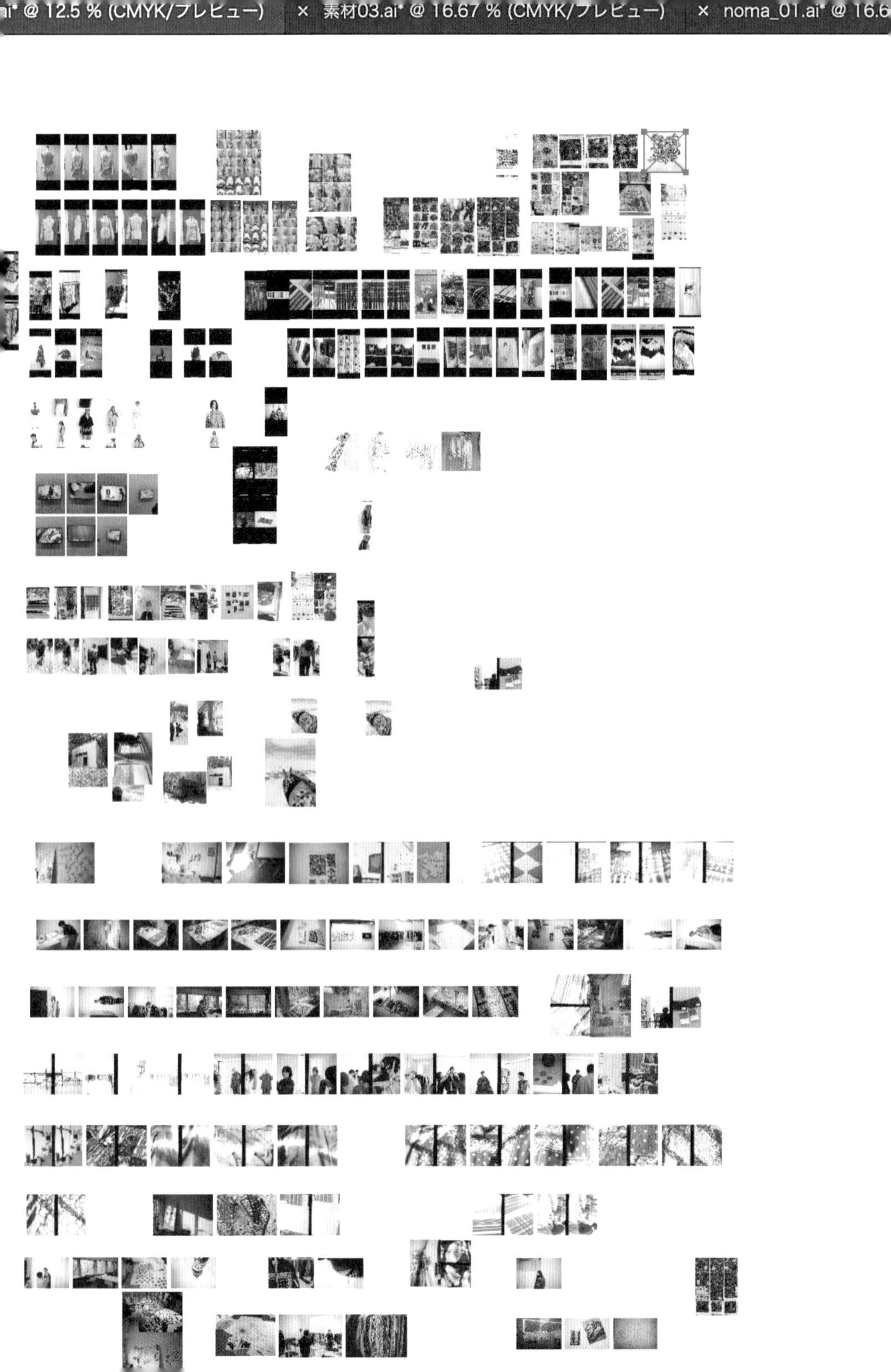
ai* @ 12.5 % (CMYK/プレビュー)
素材03.ai* @ 16.67 % (CMYK/プレビュー)
noma_01.ai* @ 16.6

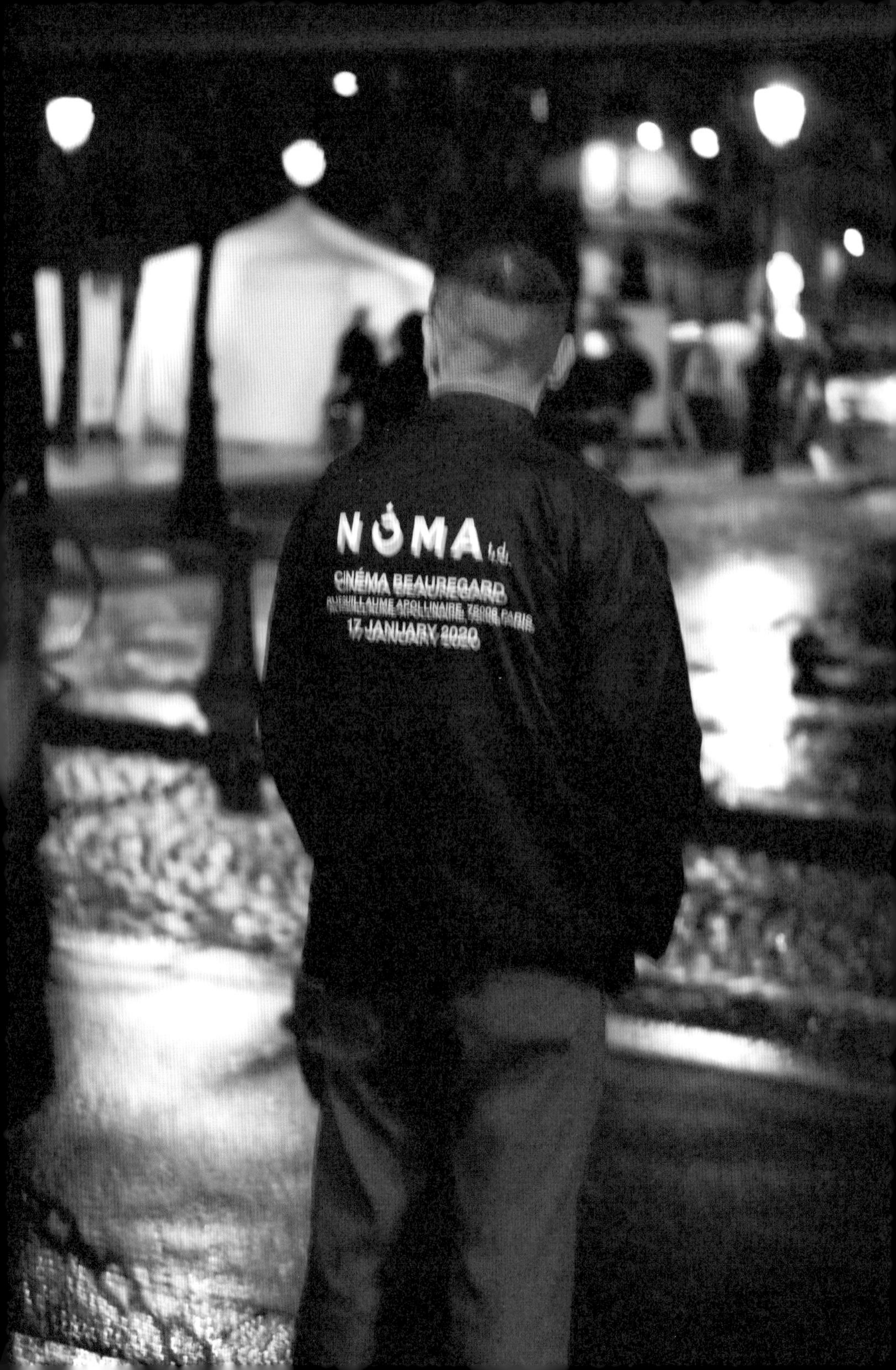

NOMA
CINÉMA BEAUREGARD
17 JANUARY 2020

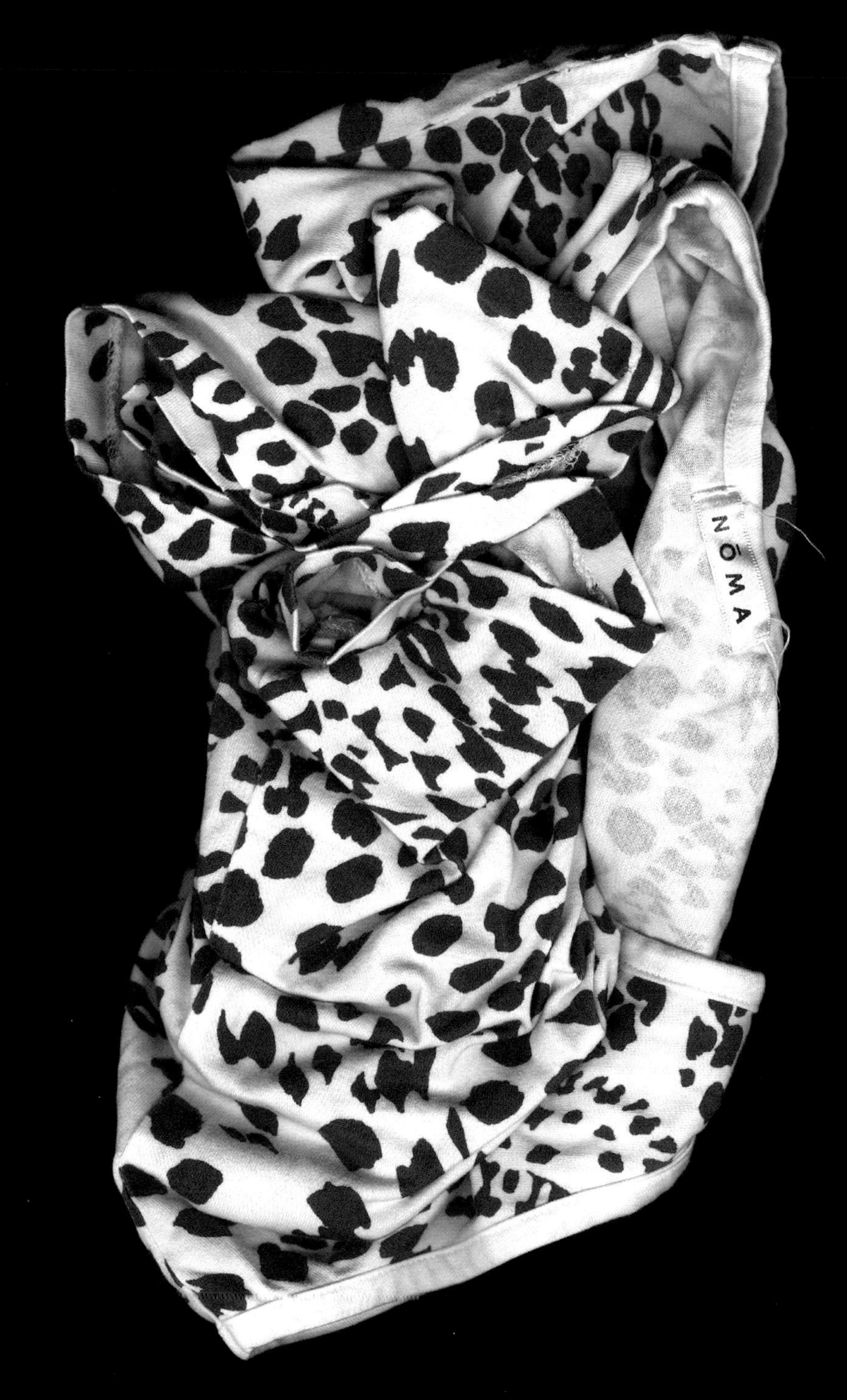
NŌMA

When I was a child, the world seemed full of lines and patterns. The flow o
flowers trembling in the breeze, the railroad tracks, electric wires, the roofs,
lines of light shines through the trees, the strong shadows that spread at my
the lines continued for eternity that the patterns expanded into infinity. I find
lines and patterns extend to the present and they fill my vision once again.¶
¶
子供の頃、世界はラインとパターンで溢れて見えた。川の流れ、風で揺れる野花、
屋根、田んぼ、木々の隙間から射す光の線、足元に広がる強い影。ラインは永遠に
ンに無限 の広がりを感じた。それらは今も、私の世界を満たしている。¶

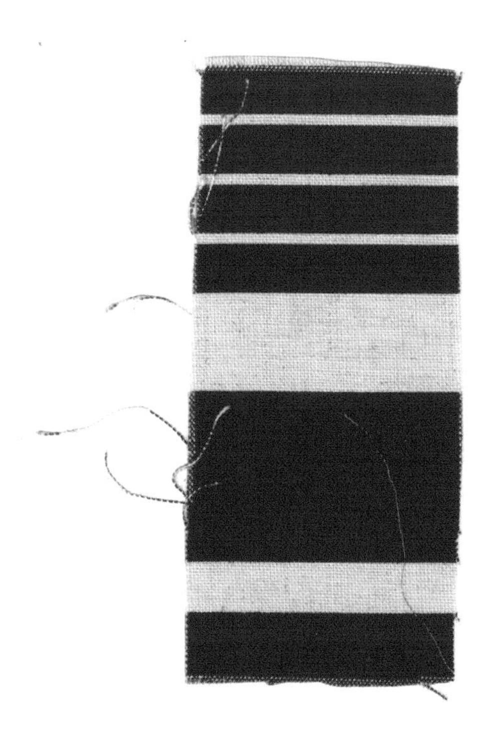

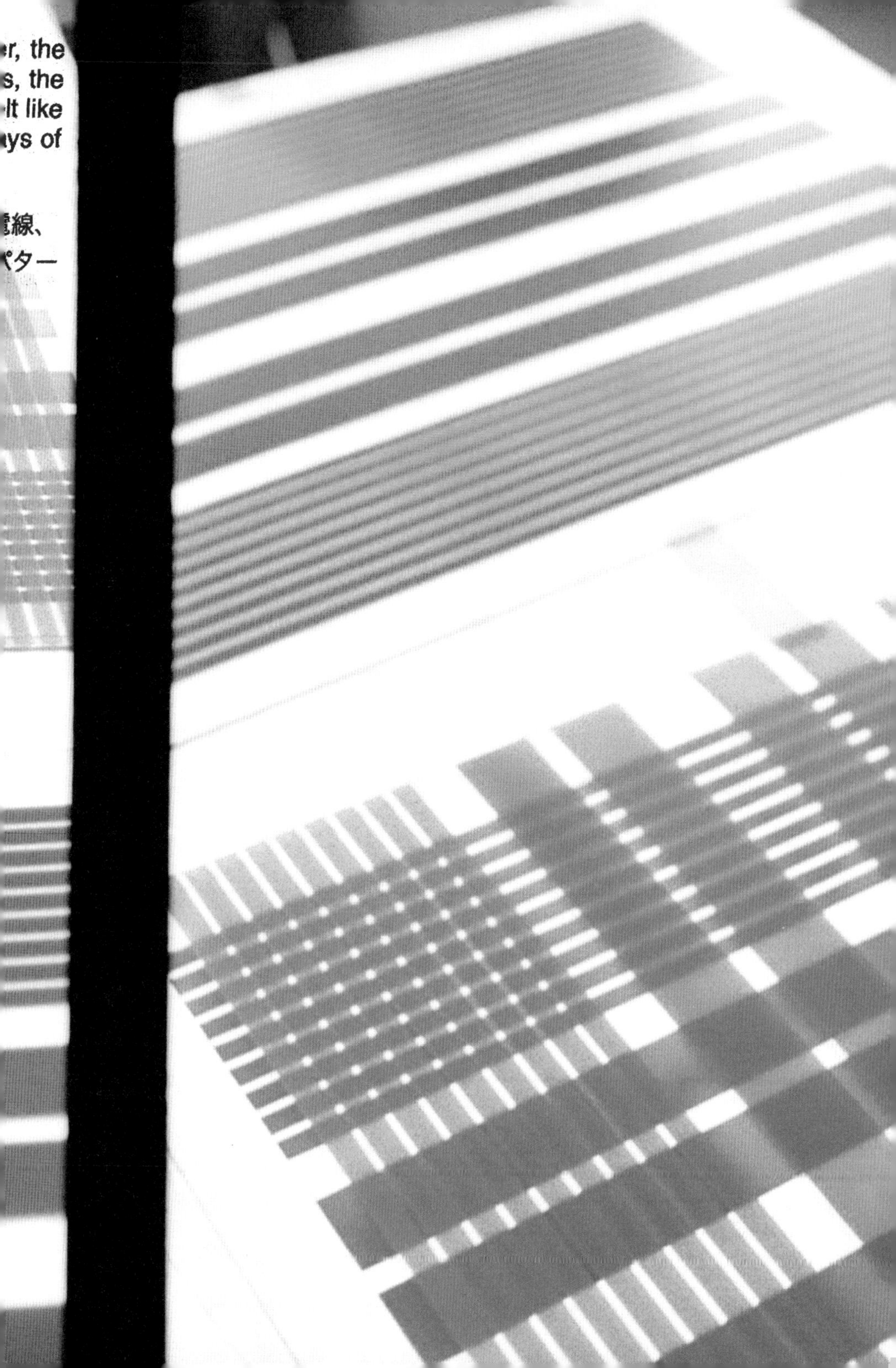

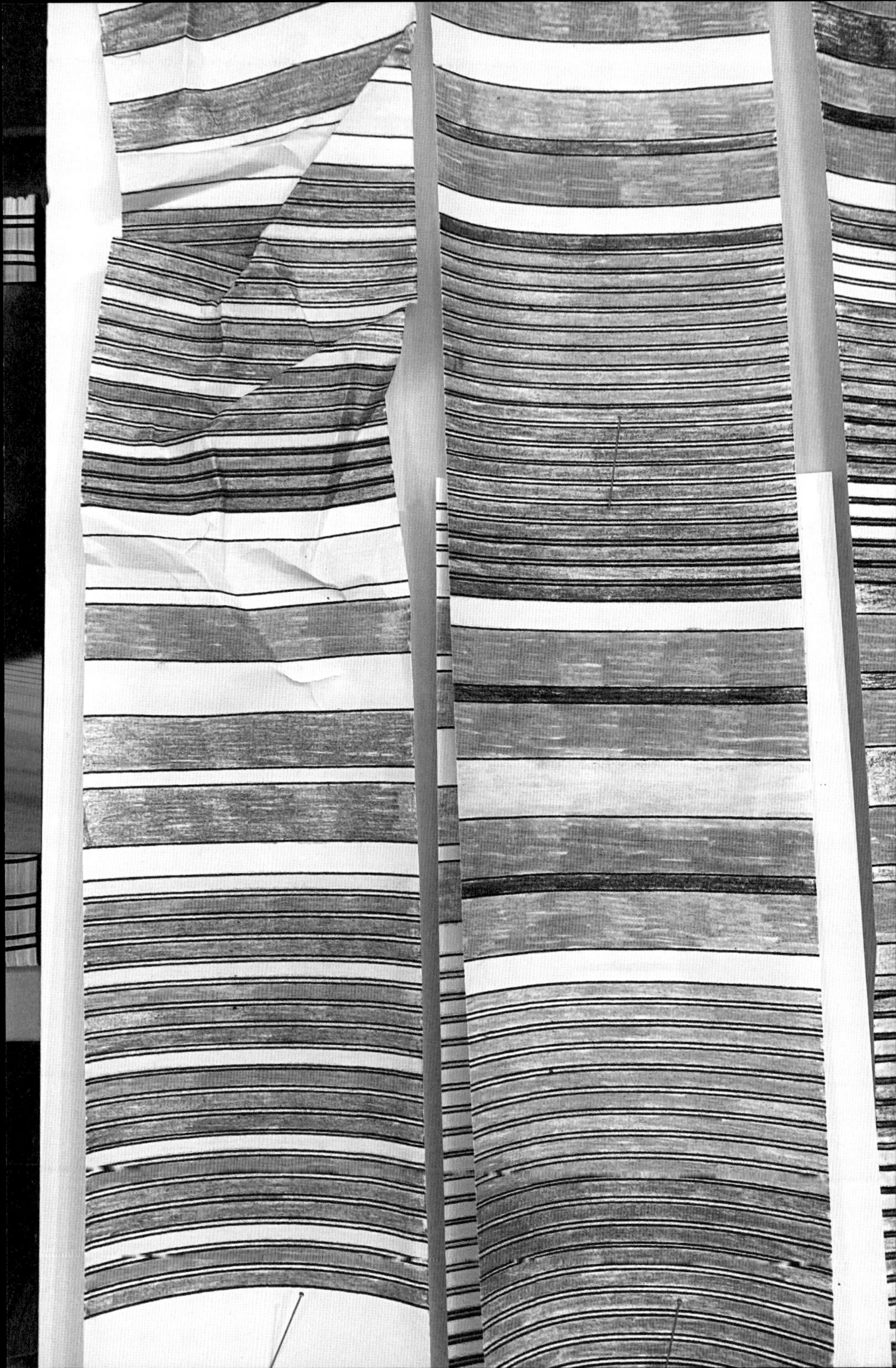

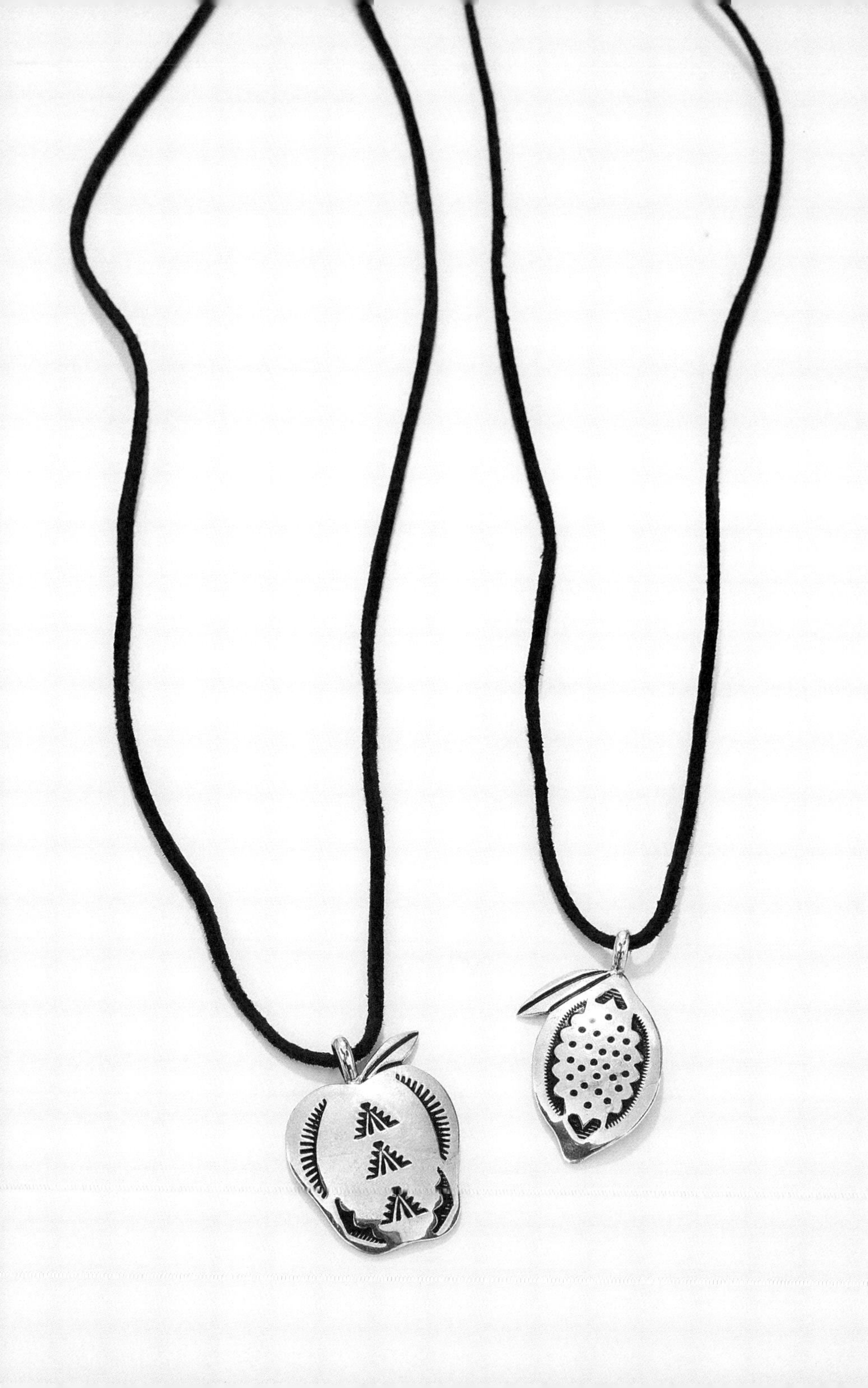

NŌMA

NOMA t.d.
#34
FEEL THE LANDSCAPE

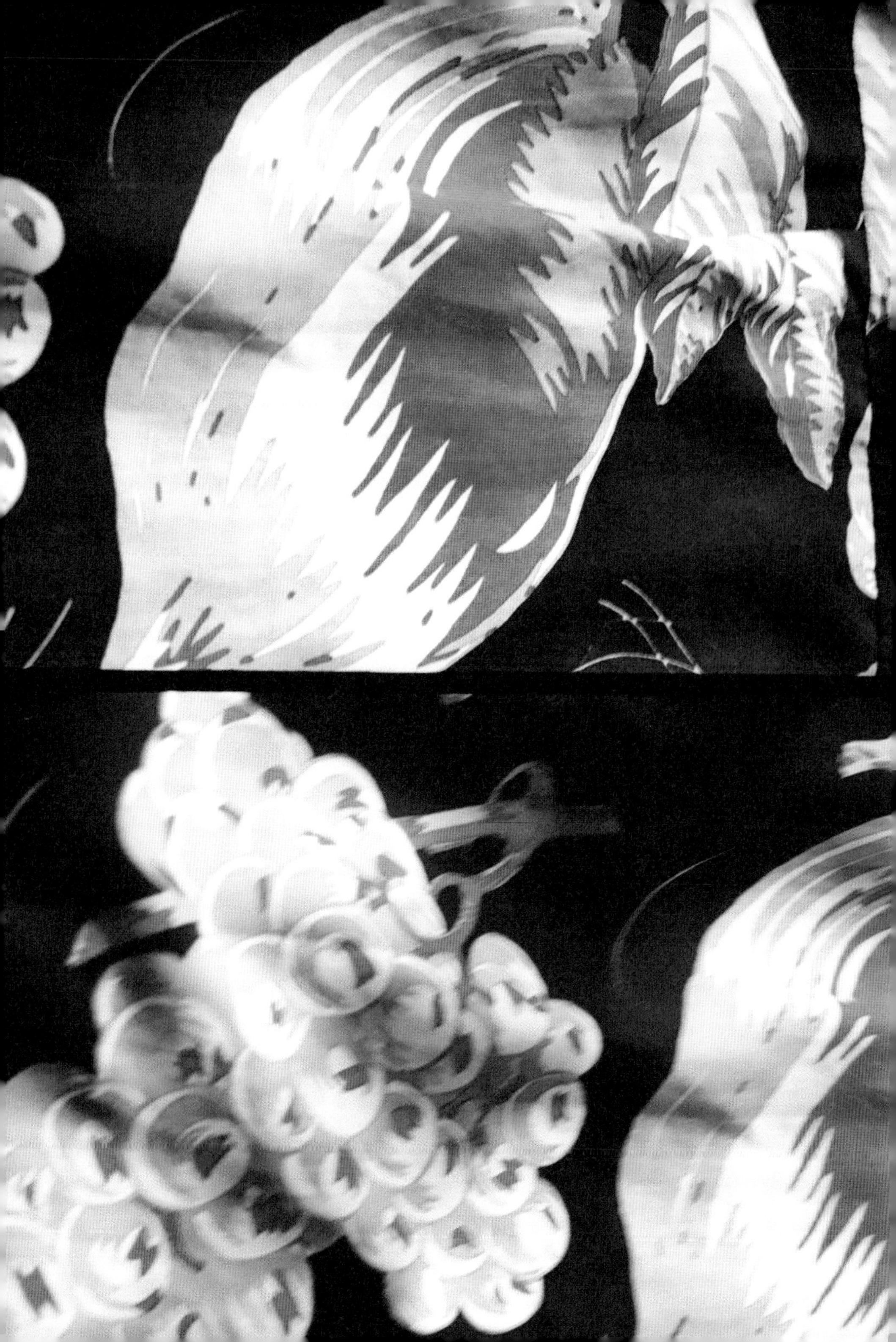

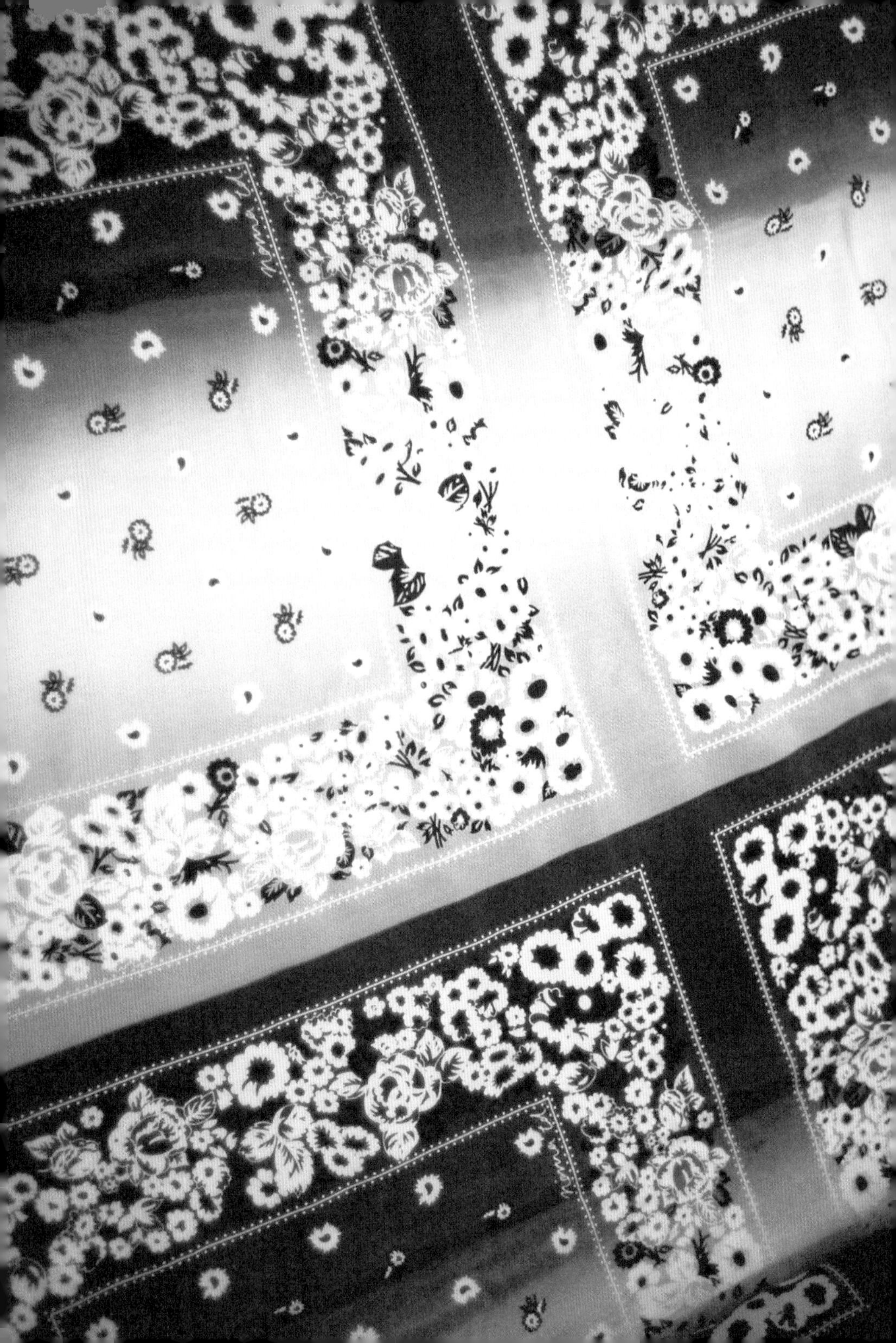

KENTUCK KNOB
FRANK LLOYD WRIGHT • ARCHITECT

NOMA.

In 2005 NOMA t.d. was launched in Tokyo by Masako Noguchi and Takuma Sasaki. We blend traditional and contemporary interpretations with varying techniques in terms of print, weave, dye, and embroidery. We regularly craft pieces by hand to achievo one-off expressions and attach aspects of individuality.¶
¶
2005年、東京にて野口真彩子と佐々木拓真によってスタート。プリント、織り、染め、刺繍などの伝統技術を、現代的解釈により融合させる試みを行なう。手作業の工程を入れ、一点ずつ表情の違う個性的なピースを制作。

Dictionary of Tools
Man Ray
BEUYS BOOK
BACK SIDE
MAKING MARKS
ANNI ALBERS
NEW ORLEANS
TASCHEN
Charlotte Perriand

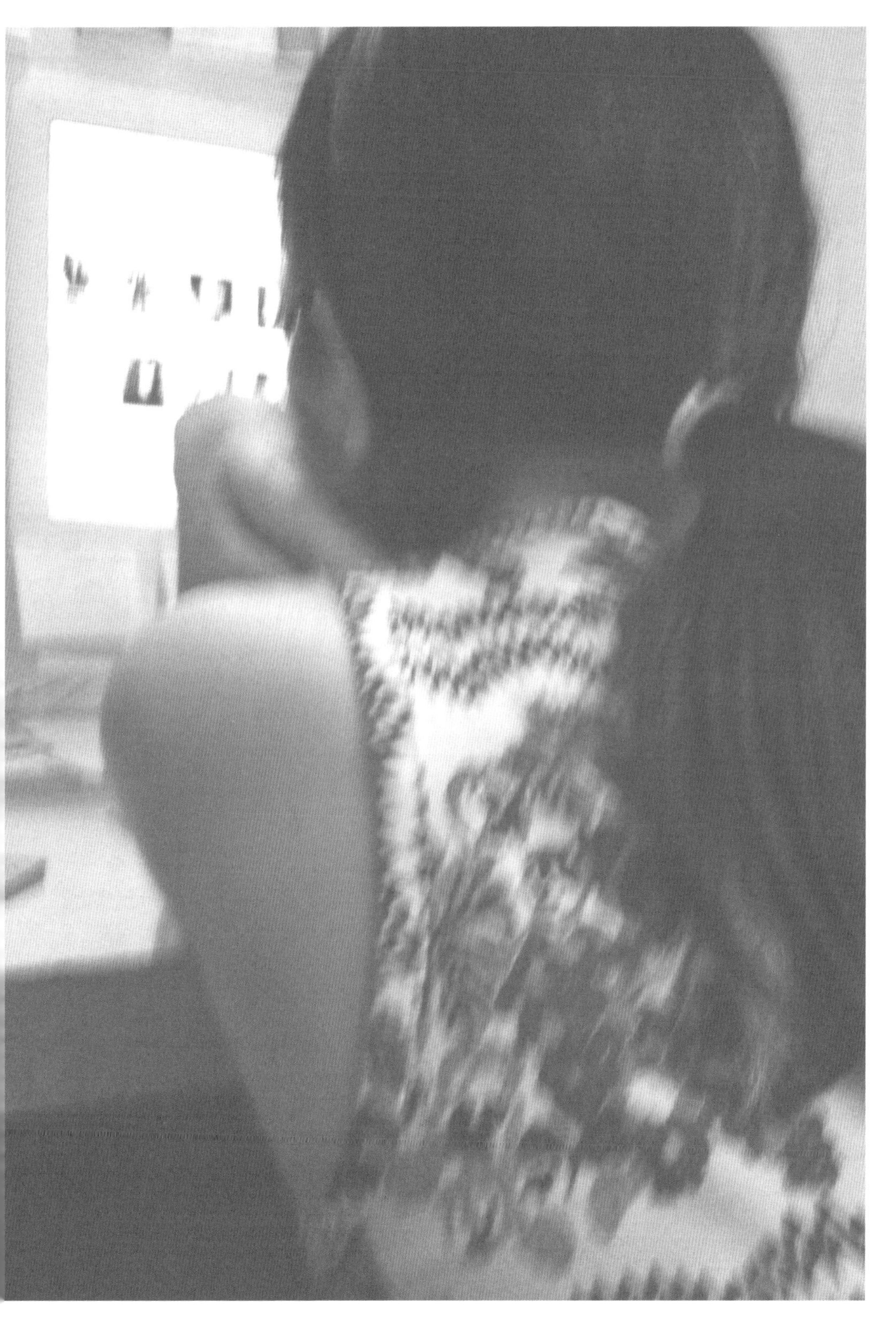

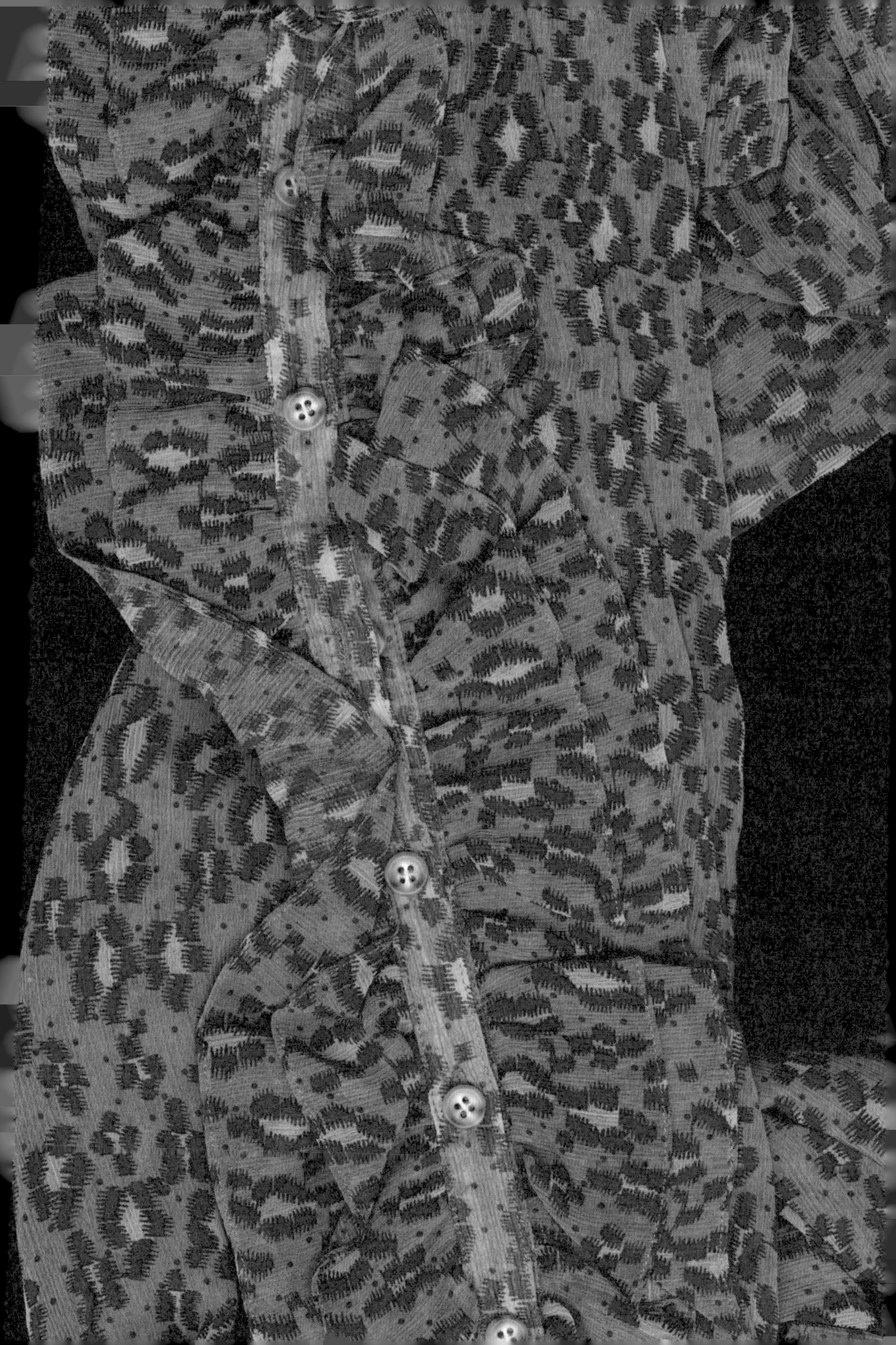

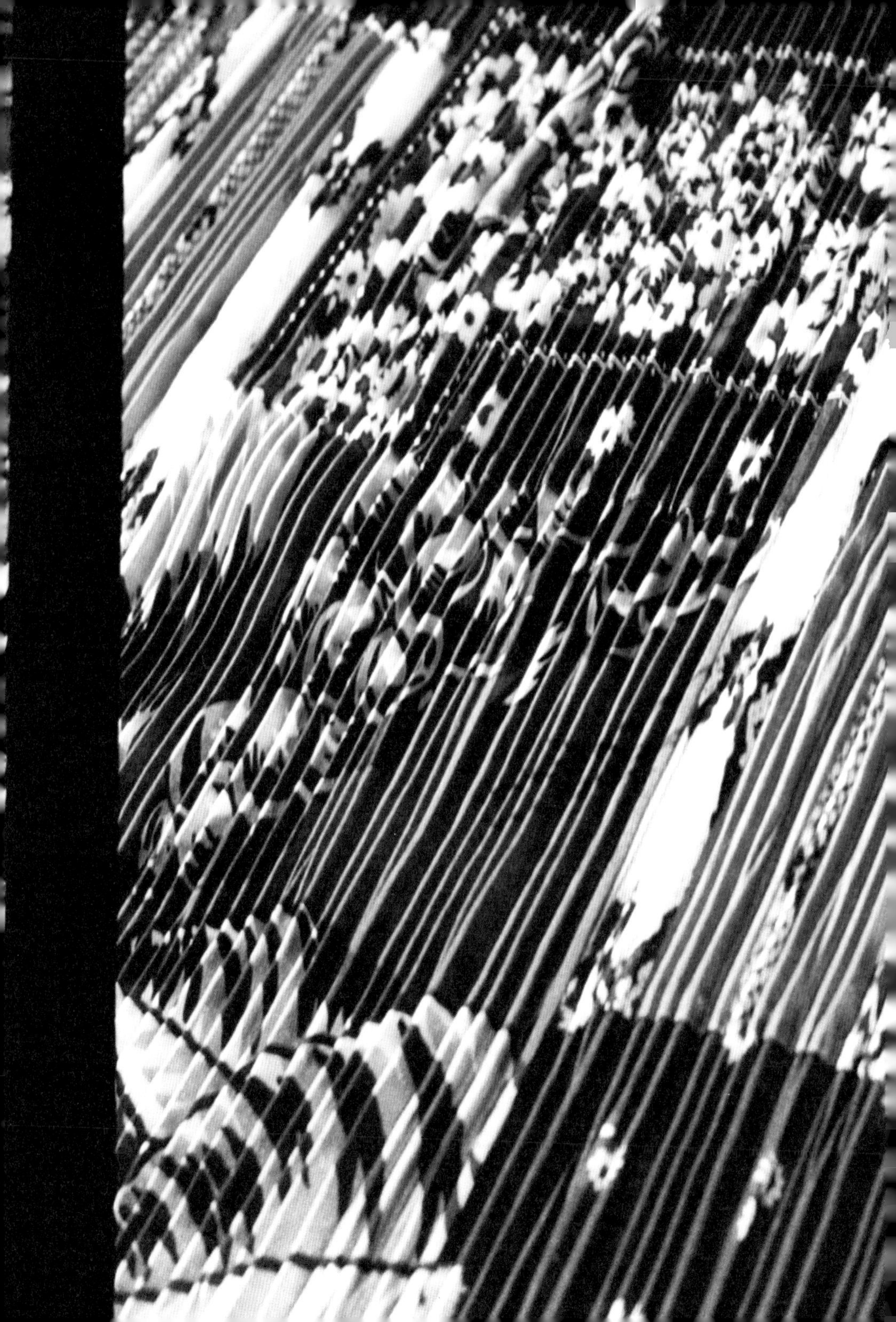

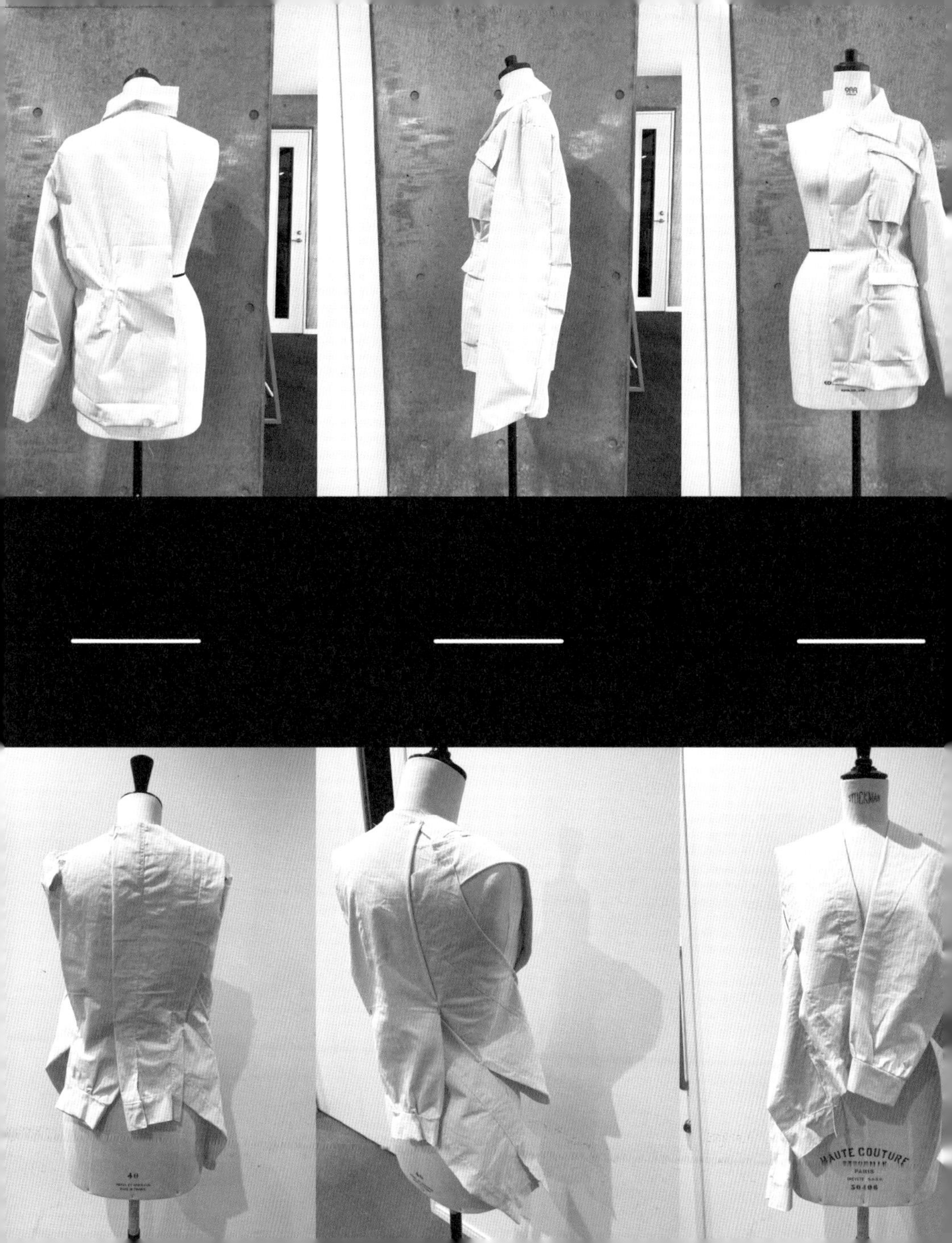
STOCKMAN
HAUTE COUTURE
PARIS

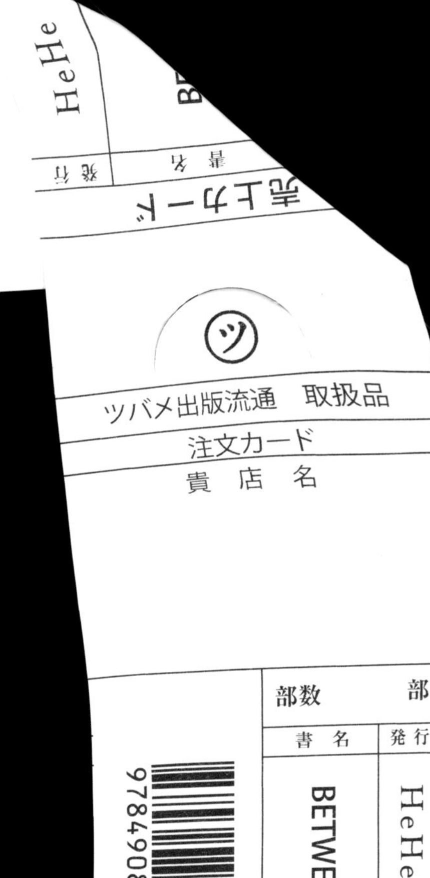

ツバメ出版流通　取扱品

注文カード

貴　店　名

部数	部
書　名	発行
BETWEEN LINE AND PATTERN	HeHe
	著者
	NOMA t.d.

9784908062483

ISBN978-4-908062-48-3

C0070 ¥2800E

本体2800円+税

ツバメ出版流通(株)

TEL 03-6715-6121

FAX 03-3721-1922

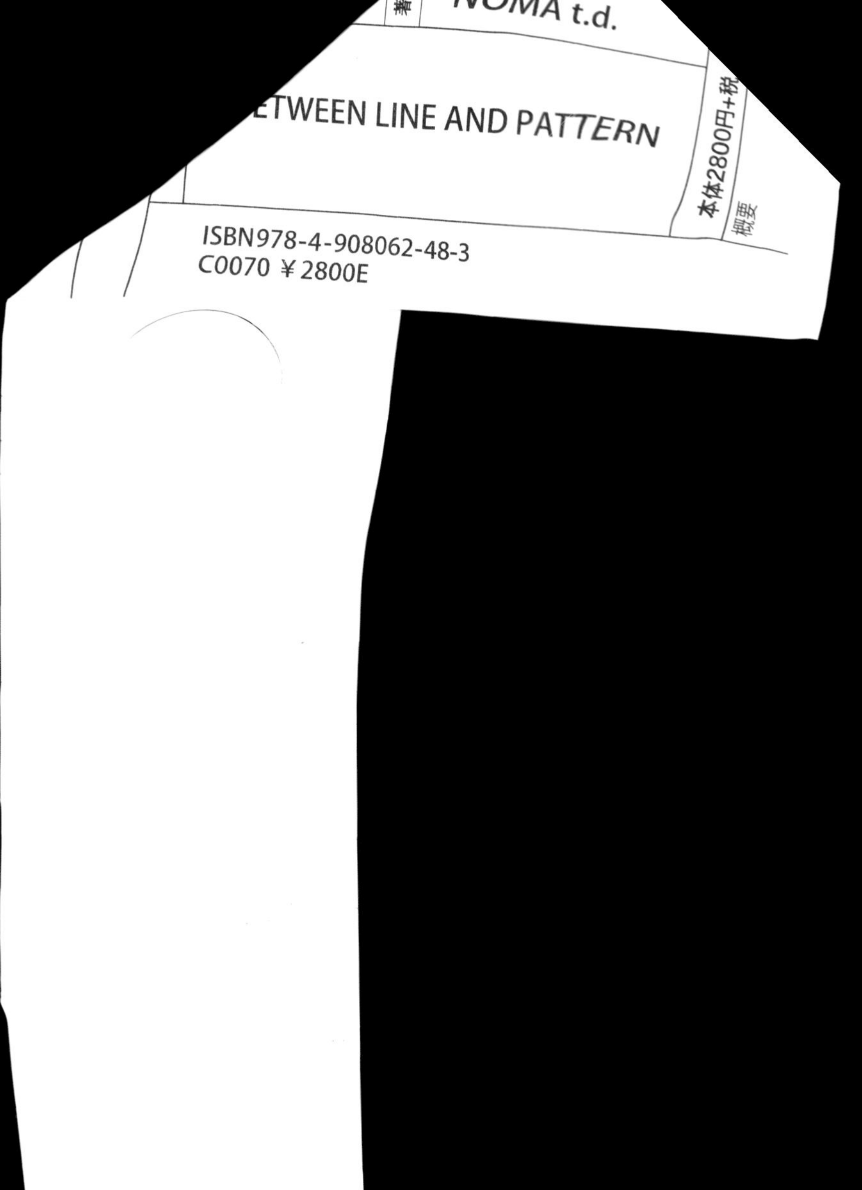

著
NOMA t.d.
ETWEEN LINE AND PATTERN
本体2800円+税
概要
ISBN978-4-908062-48-3
C0070 ¥2800E

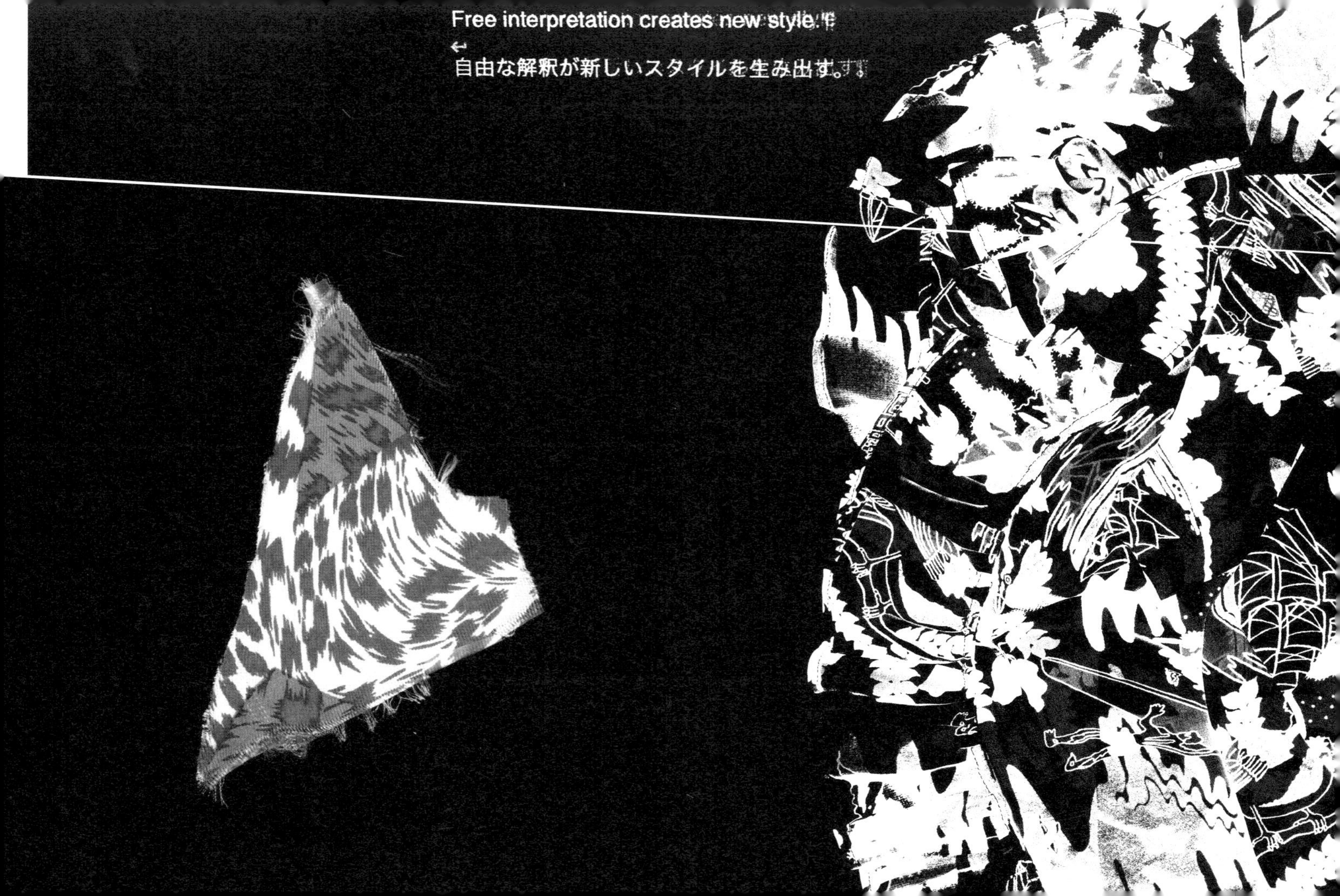
Free interpretation creates new style.
自由な解釈が新しいスタイルを生み出す。

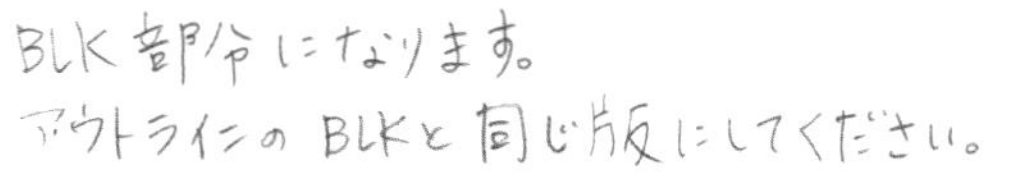
BLK部分になります。
アウトラインのBLKと同じ版にしてください。

TOSHIBA

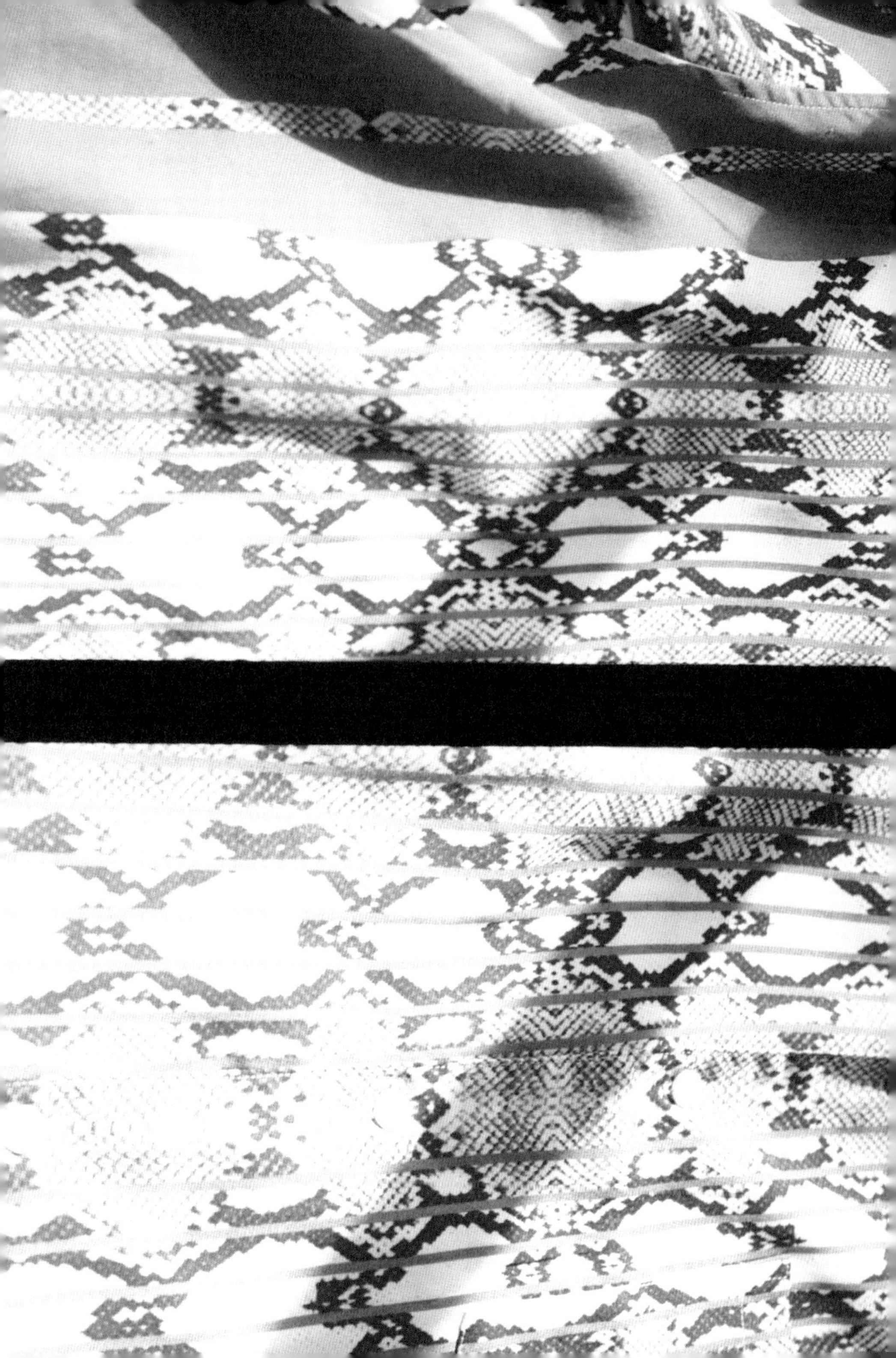

NÓMA

NOMA
HIROO TOKYO

Safari ファイル 編集 表示 履歴 ブックマーク ウインドウ ヘルプ

noma

NOMA t.d.

BETWEEN LINE AND PATTERN

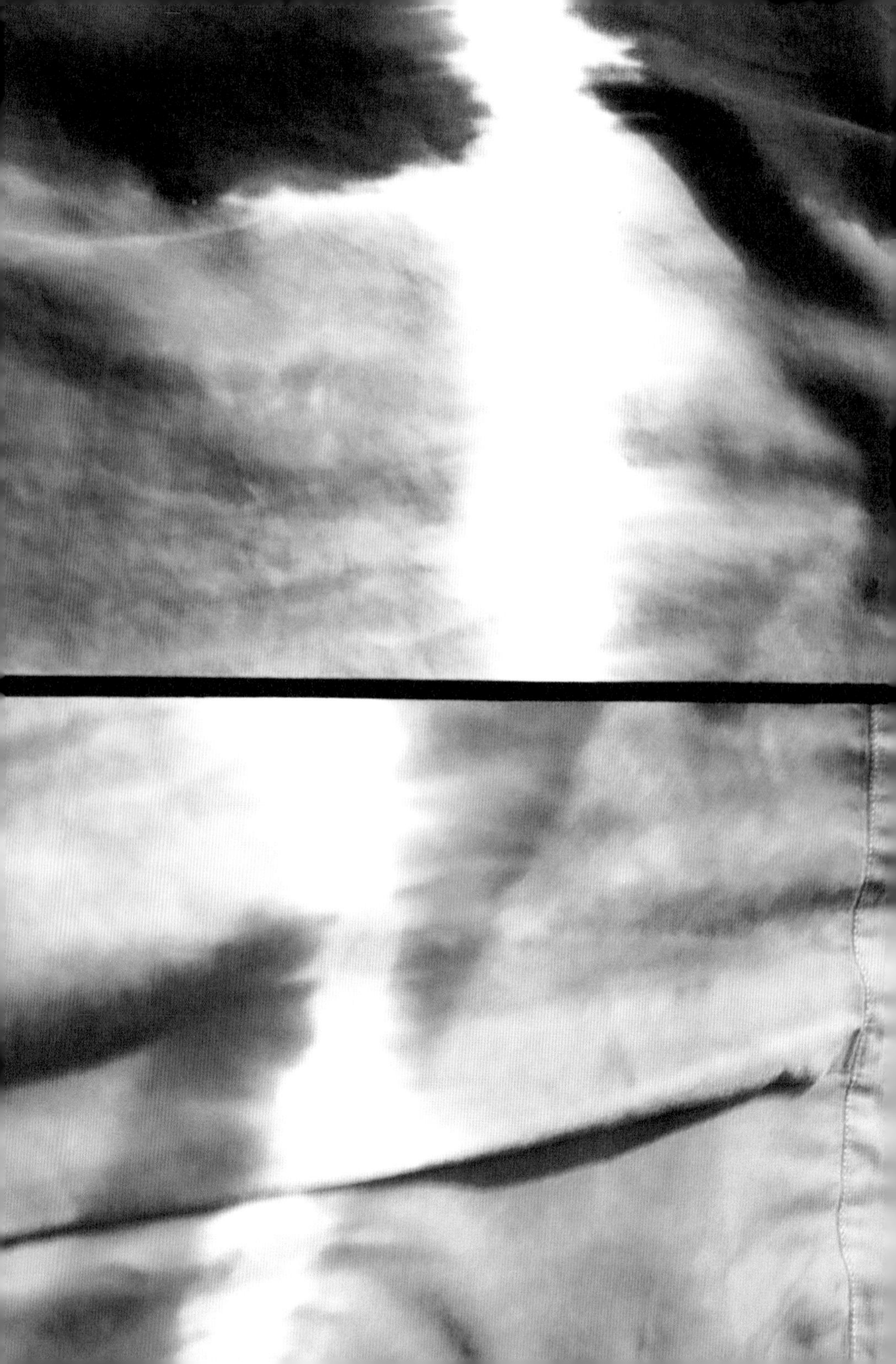

r approach is not to design textile and clothing in any particular order, but to develop h elements simultaneously. By considering the combinations, materials, silhouettes, and ails together, we believe that the best balanco between textile and garment design can¶ found.¶

キスタイルと服のデザインを順々に行うのではなく、テキスタイルのデザインと同時進行
素材、形、ディテール、違う素材との組み合わせ、縫製などを熟考する。そうすることに
り、テキスタイルと服のデザインが最も一致すると考える。

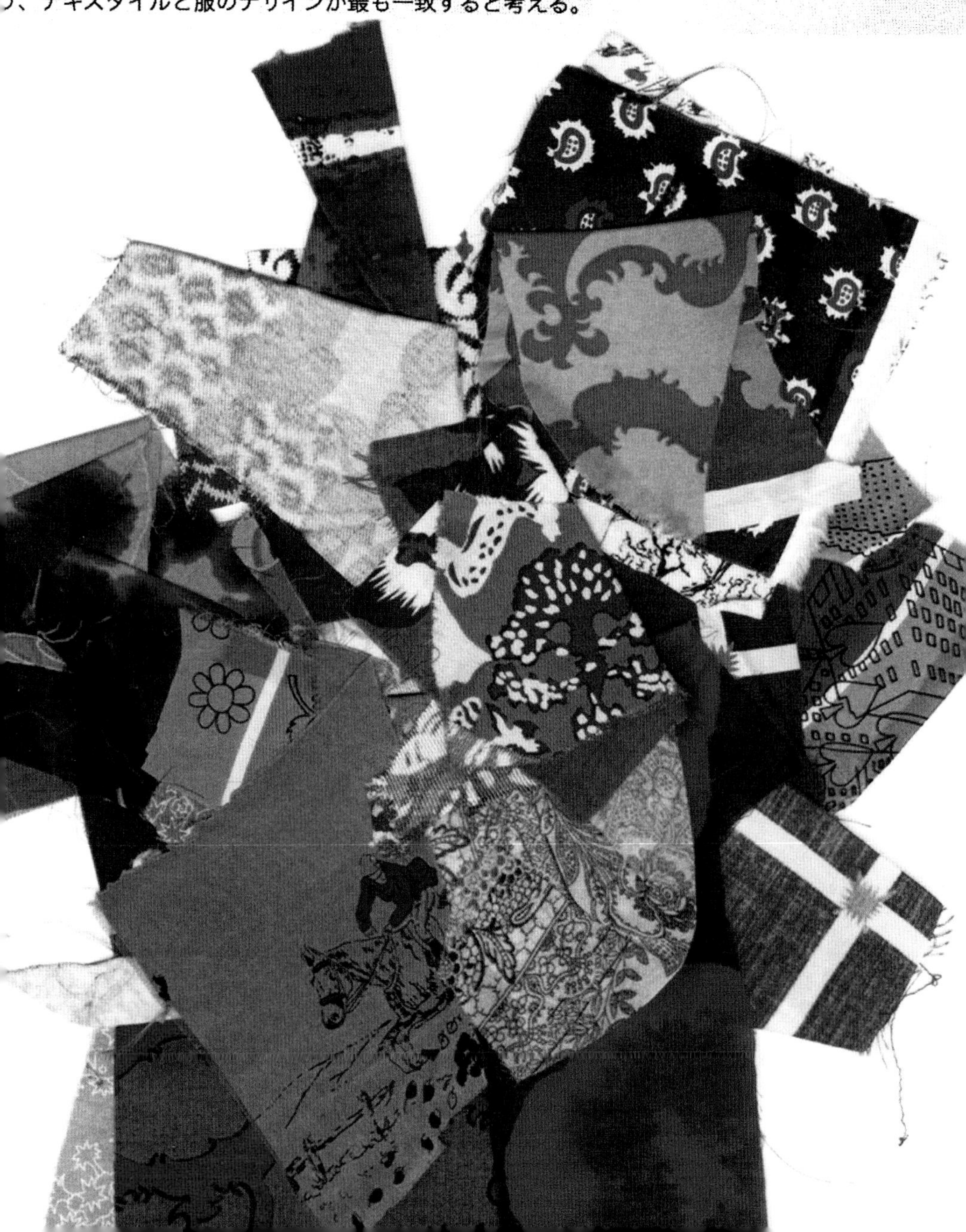

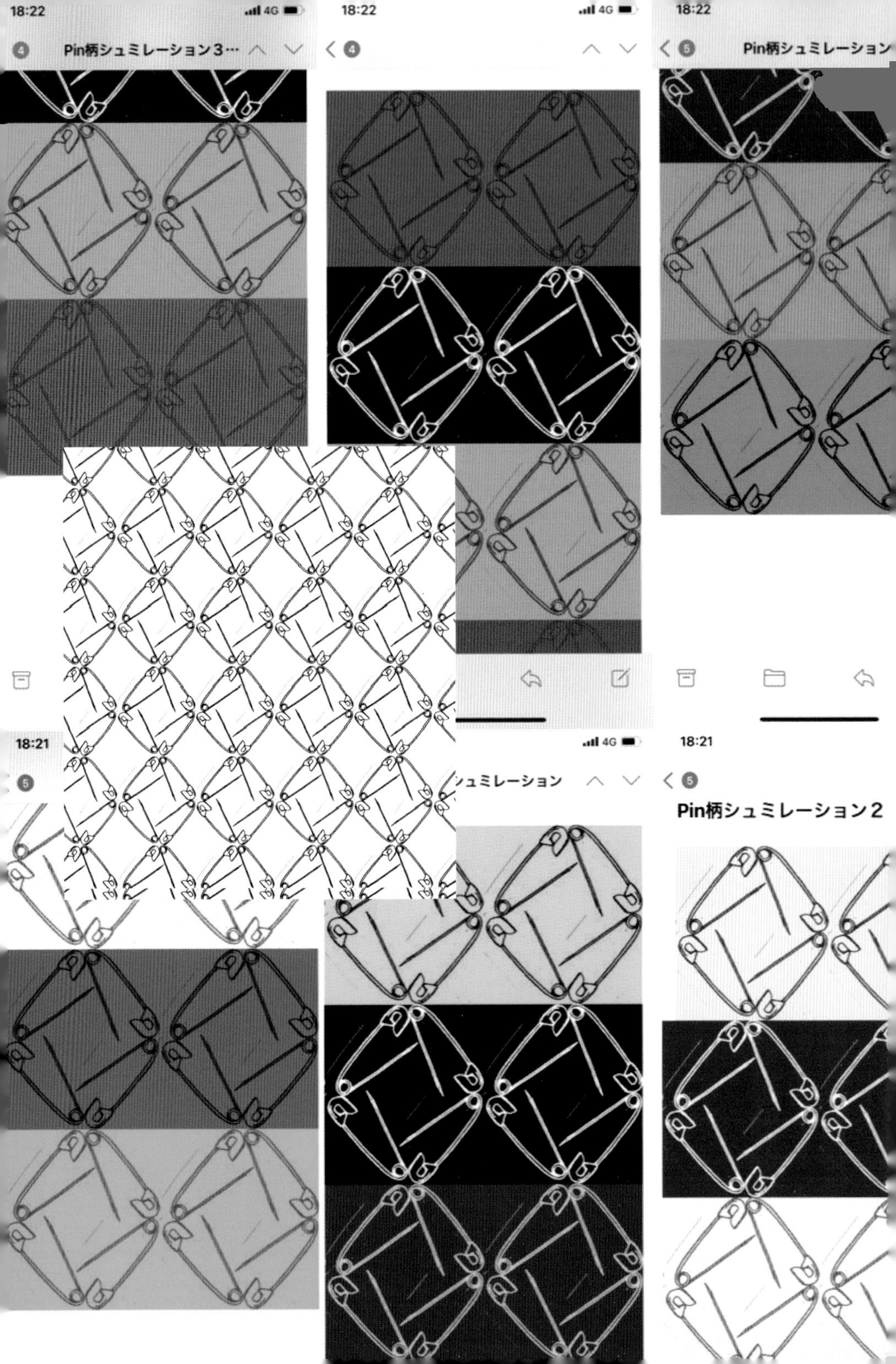
18:22
Pin柄シュミレーション3…
18:22
18:22
Pin柄シュミレーション
18:21
シュミレーション
18:21
Pin柄シュミレーション2

ACRYLIC
ACRYLIC

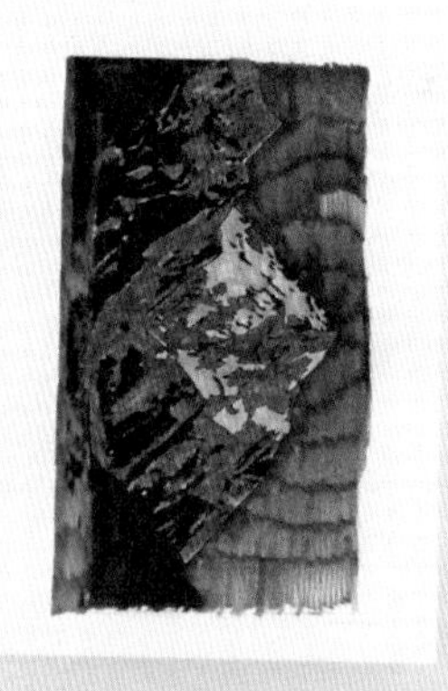

BACK LEFT | Right
OFF WHITE
BLUE